LAKE MANAGEMENT
BEST PRACTICES™

MANAGING ALGAE PROBLEMS

www.LakeAdvocates.org

Lake Advocates Publishers is the publishing arm of
Lake Advocates, a nonprofit corporation. Ordering
information for the Lake Management Best
Practices™series is available through our website.

Design: Joyce Hwang

First Printing: August 2017
ISBN 978-1-387-18446-0

Lake Advocates Publishers
5162 London Road
Duluth, MN 55804

LAKE MANAGEMENT BEST PRACTICES™

MANAGING ALGAE PROBLEMS

Dick Osgood and Harry Gibbons

LAKE ADVOCATES

MISSION

Lake Advocates facilitates scientifically based lake protection, management and restoration through applied research, policy development, training, case studies, and leadership development provided to federal, state and local agencies and organizations charged with stewarding lake health now and for the next generations. *Lake Advocates* services will be delivered using a network of experienced professionals.

VALUES

- **ACTIVE, ENGAGED COMMUNITIES.** *Lake Advocates* values communities that are engaged in the management of their lake(s) resources and take positive actions to sustain their lake(s).
- **HEALTHY LAKE ECOSYSTEMS.** *Lake Advocates* values total lake health, including a functioning ecosystem sustained by diverse biota.
- **SCIENTIFICALLY SOUND MANAGEMENT PRINCIPLES AND PRACTICES.** *Lake Advocates* values management approaches supported by sound science and practical application experience.
- **CITIZEN SELF-ADVOCACY.** *Lake Advocates* values the ability of citizens advocating for meaningful and lasting changes in policy and practice at appropriate levels - by property owners, individuals, lake users, developers, policymakers, and others.

SERVICES

- Client, Community Services
- Agency Support – Technical, Applied Research & Education
- Policy Analysis / Emerging Issues
- Training: Limnological, Management Technologies
- Advocacy
- Product and Services Evaluation
- Program Reviews
- Publications
- Mediation

www.LakeAdvocates.org

ACKNOWLEDGMENTS

Several individuals, listed below, have made valuable suggests, which have improved this book. We thank:

Dick Fowler
Joan Hardy
Mark Hoyer
Joyce Hwang
Jean Jacoby
Steve Lundt
David Rosenthal
Ken Wagner

Ann St. Amand, owner of PhycoTech, Inc. and *Lake Advocates*' Strategic Partner, provided photos of algae.

CONTENTS

PREFACE

We have been involved in the profession of lake management since it was first recognized. In the early years, professionals, like us, were educated in the field of limnology – the science of lakes. Early on, lake management was spawned from an outcry to restore lakes impacted by sewage.

We remember the headline in 1969, "Is Lake Erie Dead?" With this call to action, the Clean Water Act (1972) got down to the business of restoring lakes, mainly by diverting or treating sewage effluents. These early efforts were well funded and largely successful. Many lakes got better.

As the practice of lake restoration evolved and many of the high-profile lakes were improved, the approach to managing the more numerous, but less known, impacted lakes faced new challenges. It was obvious that sewage was harmful for lakes, but it was not so obvious that pollution also came from other sources, such as urban and agricultural runoff. These sources were more challenging to define and even more challenging to mitigate.

Today, funding for and the administration of lake management has been diminished and decentralized, leaving local communities, water management districts, lake associations and others taking the lead for managing their lakes. Although lake management techniques and technologies have evolved and improved overall, some of these tools have not proved to be reliable or effective.

If your lake has water quality problems, how can you effectively steward and advocate to find the best remedy for your lake?

This book is designed to help answer that question, as a first step in the right direction[1].

Using our combined professional experience of nearly a century as well as scientifically based literature and objective assessments, we can simplify and clarify lake management best practices.

To keep this book short and readable, we do not include specific reference citations or other basis material. Be assured though, that our assessment is based on many years' experience, our understanding of the literature and numerous case studies. We expect and encourage an ongoing dialog. We will address readers' questions and concerns through our blog (www.LakeAdvocates.org) and provide clarifications, updates, references and specific answers on an ongoing basis. Thus, readers will receive the most up-to-date, relevant and useful information. In this way, we won't be so bold as to claim this book's timelessness, but we hope it comes close.

Please note that we do not have financial interests in any one management alternative. As co-founders of **Lake Advocates**, we are independent limnologists and lake managers, working to promote the sound science of lake management that is sustainable and works in appropriate situations.

---

[1] Significant portions of this book are taken from *"Do you want something that works?"* Dick Osgood, LakeLine. Spring 2015:8-16 and used with permission of the North American Lake Management Society.

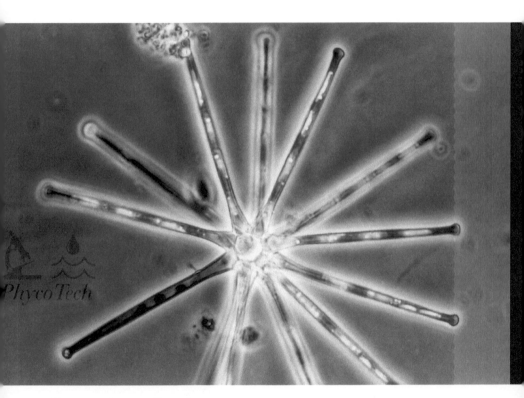

Figure 1. Asrerionella, a diatom.

INTRODUCTION

DO YOU WANT
SOMETHING THAT WORKS?

This question is too seldom asked when applied to lake management. The modern profession of lake management has been around for about 70 years and has evolved to a point where we have an obligation to be critical, systematic and scientific in our approach. There are many tools, techniques and approaches that: a) work well but are not always used where appropriate or b) are unproven, yet are used often without defensible results. Here we review lake management approaches that work with measurable outcomes and lake management approaches that are not ready for prime time due to insufficient development or inadequate testing.

> Lake management techniques or approaches that "are not ready for prime time" doesn't mean they will never be or should be avoided, rather there must be an understanding that the application entails uncertainty and careful monitoring is recommended to evaluate outcomes.

Seldom does anyone ask us "what works?" Those involved in managing lakes have become less demanding of outcomes and more concerned about what are the acceptable, correct, popular or expedient methods. Our lake management institutions are charged with various aspects of managing issues (through regulations, funding, education, etc.), but not resolving them – a serious flaw in the system.

For the most part, we have the tools to manage lakes for positive outcomes. Our concern is that many lake managers as well as clients and stakeholders too often set aside these tools for approaches that appear preferable. Lake management feasibility should involve assessments of applicability and reliability first, followed by evaluation of costs and regulatory acceptability. And, we must seek and expect real physical and ecological outcomes in the lakes we manage.

Here we provide a screening tool for those confronting lake management challenges concerning algae problems.

STATUS QUO

Some lake managers and those we serve seem to want:

- What is politically acceptable
- What is popular
- Magical or mysterious properties pills
- Quick fixes
- Natural remedies
- Non-chemical remedies
- Cheap or affordable remedies

However, we seem to not want or to know what works and what does not work, or even how best to facilitate or evaluate success. We have become lax on standards of care and due diligence as well as proper planning principles.

**This book is designed to know
what works and how well it works.**

The result is that we too often lack positive outcomes, we have engendered a proliferation of tools, techniques and approaches that are untested or unreliable, and we have set aside tried and true methods that do work, even though they sometimes offend certain lake users or regulators for various reasons that may not hold up under scrutiny.

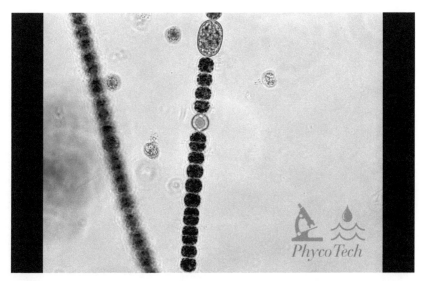

Figure 2. Dolichospherum, a bluegreen algae (or cyanobacteria).

MANAGING ALGAE PROBLEMS

This book is divided into two sections: Algae Problems Overview and Best Practices.

The Overview briefly describes algae problems, their cause and categorical management approaches. Algae Management Best Practices provides a summary of best practices and evaluates whether they "work" or not based on their applicability and reliability. Our evaluation relies on the high scientific standard of peer-reviewed journal citations or objective, third-party assessments.

There are plants in lakes other than algae. Specifically, rooted plants anchored to shallow lake bottoms. These plants, sometimes called weeds, are often problematic and a frequent target for management actions. Managing weed problems will be covered in another book in the Lake Management Best Practices™ series.

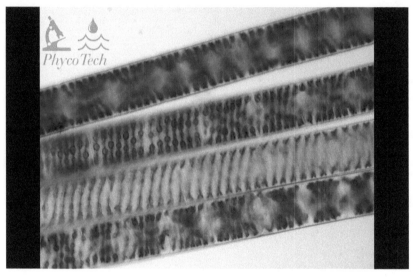

Figure 3. Spirogyra, a green algae - note the spiral chloroplasts.

OVERVIEW: ALGAE PROBLEMS

Algae are microscopic plants (sometimes long filaments) that are an essential and natural part of lakes, representing the base of the food chain in healthy lakes. Unfortunately, sometimes algae grow too abundantly, form unsightly surface scums, result in oxygen loss or produce toxins. Algae are problematic in these cases.

Algae problems represent an imbalance. There are cases where a great deal of algae may be produced, but they may be quickly consumed or settle and therefore do not accumulate. More commonly, excess phosphorus (P) tips this balance – then algae problems occur.

HABs—or Harmful Algal Blooms— are on the rise.

Exposure to waters with **HABs** is a serious public health concern because of the many different potential types of toxins that can be produce during a HAB event.

HABs are most commonly comprised of blue-green algae, also known as cyanobacteria or blue-green bacteria.

WHAT CAUSES ALGAE PROBLEMS?

Excess nutrient availability in lake water results in overabundant production of algae. The nutrients of concern have been carbon, nitrogen and phosphorus. However, it has been amply demonstrated that only the control of one of these nutrients, P, has yielded consistent results relative to limiting the occurrence and intensity of algal problems[2]. Phosphorus availability in the lake must be controlled to mitigate algae problems and regain the usability of the lake.

PHOSPHORUS. Phosphorus is the element in lake water in shortest supply relative to the growth needs of algae. While there are sometimes exceptions, as a practical matter, P is the key factor controlling algae growth. The more P in the water, the more algae will grow (and the more obnoxious, problematic kinds of algae too).

Phosphorus is measured and reported based on its various forms found in water. Total phosphorus is the sum of all chemical and biological forms of P. Because P is readily transformed from one form to another, total P is the best single indicator. Total P is the sum of particulate P and dissolved P.

---

[2] Nitrogen may also be limiting or co-limiting algae problems. Usually, low nitrogen-to-phosphorus ratios (in the water) are indicative of nitrogen limitation. However, in many cases, this also means there is excess P, which will need to be mitigated.

PHOSPHORUS SOURCES & FATE

The amount of P in lake water is the remainder of P that enters a lake and the amount that leaves the lake or settles to the lake bottom. P sources are often enhanced or increased by human activities, such as land development, yard litter, fertilizer application and agricultural activities.

P SOURCES

- Atmospheric – rain, snow and windblown
- Runoff – surface drainage
- Discharges – municipal or industrial waste
- Groundwater – seeping into the lake bed
- Internal – recycled from sediment source and/or from in-lake biota

P LOSSES

- Lake outflow
- Groundwater - seeping out
- Withdrawals
- Sedimentation – settling to the lake bottom
  (note some of this will come back as internal P sources)

Algae live in and respond to the amount of P in the lake's surface water. Inputs and losses may occur at the surface as well as between the surface (top) and bottom layers (as delineated by a thermocline) and between the bottom waters and the lake's sediments.

Reducing lake surface P – which is where algae use it – involves reducing inputs, accelerating outlets, or both.

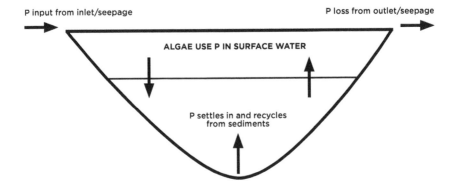

P input from inlet/seepage   P loss from outlet/seepage

ALGAE USE P IN SURFACE WATER

P settles in and recycles
from sediments

*Figure 4. Sources and fates of phosphorus in lakes.
Note, this is a simplification.*

Of special concern for algae problems is the internal recycling of P, because this P is usually available during the algal growth season. Internal recycling occurs following a period of excess P inputs. Most of the P that enters a lake stays in the basin (commonly 90% or more) and therefore builds up in lake sediments. As lakes become eutrophic, this accumulated sediment P will be released back into the lake water. When internal P cycling occurs, it slows lake recovery.

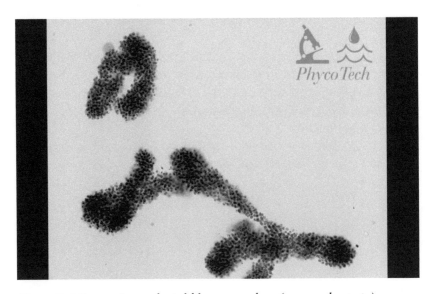

Figure 5. Microcystis, a colonial blue-green algae (or cyanobacteria).

MANAGEMENT APPROACHES

We emphasize that the first step to management is to abide by sound planning principles.

PLANNING PRINCIPLES

- **Define the problem**
- **Identify the source or cause**
- **Define a measurable management objective**
- **Evaluate feasible management alternatives**
- **Identify a sustainable funding source**
- **Have a lead management authority**
- **Implement the plan**
- **Monitor progress**
- **Adjust – adapt as needed based on data and lake needs**

MANAGEMENT APPROACHES FOR ALGAE PROBLEMS

Of greatest import is to have clear, meaningful and measurable objectives. For algae problems, typical objectives include less algae overall, lower frequency of algae blooms, decreased prevalence or occurrence of HABs or increased water clarity. If P is the cause of the algae problem, a management objective that includes reducing P is appropriate. If P controls are not practical or feasible, then a management objective directly relating to the algae problem is appropriate.

Below we describe three management approaches for reference. In the following table, specific management techniques are categorized with respect to which approach they are designed to address. We emphasize that the decision to use a particular technique or approach must follow appropriate diagnostic evaluations to assure its appropriate (and effective) application.

1. CONTROL THE SOURCE OF PHOSPHORUS FROM THE WATERSHED. Municipal (sewage) and industrial discharges are highly regulated, but still, even treated wastewater can be a major P source in lakes. Of the remaining P sources noted above, P in runoff is the most common offending source. To the extent practical, this source (if identified as significant) should be mitigated.

2. CONTROLLING THE INTERNAL SOURCE OF P. When internal P recycling is a significant source, mitigating it is appropriate. In many cases, internal P will need mitigation.

3. DIRECTLY CONTROLLING ALGAE. Direct algae controls are sometimes referred to as controlling the symptoms rather than the problem or cause of lake impairment. This is accurate for the most part, but can be justified, especially if P source controls are not feasible.

4. NO CONTROL. Intentional lack of control will not be helpful, but it is also true that some technologies have no proven record (at least their reliability is very low and should not be relied upon at this time) for effectively mitigating algae problems.

Specific algae control techniques are presented, evaluated and discussed in the next chapter. The table below summarizes the categorical management approach appropriate for each.

	WATERSHED P CONTROL	INTERNAL P CONTROL	ALGAE CONTROL	NO CONTROL
Algaecides			💧	
Artificial Circulation		💧	💧	
Biomanipulation			💧	
Drawdown		💧		
Dredging		💧		
Microbes/Enzymes				💧
Oxygenation		💧		
Public Education	?[3]			
Watershed Management	💧			
P Precipitation: alum	💧	💧	💧	
P Precipitation: Ca, Fe	💧	💧		
P Precipitation: other	💧			

[3] Public education is often misunderstood in regard to measurable outcomes. See the Public Education section in the next chapter for further explanation.

A WORD ABOUT THE ECOLOGICAL AND REGULATORY LINGO

Lakes occur in two recognized states relative to their P content: oligotrophic and eutrophic. Oligotrophic means nutrient-poor and eutrophic means nutrient-rich. Mesotrophic is in between.

As the P concentration increases, lakes' biological productivity increases because P fuels lakes' food chains. Unfortunately, there is a limit to the beneficial aspects of this productivity. At some point, too much biological activity – particularly algae production – becomes problematic.

There is a continuum between oligotrophic and eutrophic. As a practical matter, we assess this by measuring total P concentration in the lake's surface water. There are no hard and fast thresholds between states – there are several intermediate states that can be relevant to algae problems. Careful diagnostic studies are therefore critical to effectively managing algae problems.

EQUIVALENT P CONCENTRATION UNITS[4]

- Micrograms per liter (µg/L)
- Milligrams per cubic meter (mg/m3)
- Parts per billion (ppb)

Because P fluctuates – daily & seasonally – the most useful measure is a seasonal average.

A common demarcation between oligotrophic and eutrophic is a total P concentration of 30 ppb, although this varies in some regions (ranging from as low as 10 to as high as 60 ppb). Hence, mesotrophic, which has a total P concentration between 12 and

---

[4] P concentrations are very small amounts. For example, 20 ppb represents about 20 drops of water in an Olympic-size swimming pool.

40 ppb, yet can still have episodic algal blooms. In practical terms, above this point, algae problems begin to develop and intensify as P concentration increases.

In the context of the Clean Water Act, states are required to designate lake impairments (by cause or source of impairment), then develop plans to remedy the impairments. For each impairment source, such as P, standards are developed to indicate whether an impairment exists. Phosphorus impairments are generally synonymous with being eutrophic, although specific P standards vary.

BEST PRACTICES

SCREENING TOOL

We present a screening tool for whether or to what extent a particular lake management approach or technique works for managing algae problems. We use a two-tiered approach based on applicability and reliability.

Our evaluation is based on our knowledge of the literature and extensive field experience.

APPLICABILITY

Is this technique, approach or method applicable?

- **HIGH.** Applicable to the stated problem(s) to be managed, mechanism or mode of action is understood, risks are well known and minimal with proper application and implementation.
- **MEDIUM.** In between "high" and "low."
- **LOW.** Not applicable to the stated problem(s), unknown mechanism or mode of action, non-target impacts likely or unacceptable level of risks.

RELIABILITY

What is the efficacy, reliability and repeatability of technique, approach or method?

- **HIGH.** Efficacy, reliability and repeatability well established in peer-reviewed literature relative to the science that supports its applicability. Track record demonstrates positive outcomes with proper application and implementation.
- **MEDIUM.** Efficacy, reliability and repeatability established in a sufficient number of case studies evaluated by third-party, independent assessments or reviews using pre- and post-data to indicate a likelihood of positive outcomes.
- **LOW.** Efficacy, reliability and repeatability not clearly demonstrated or only claimed by potentially biased sources. Additional assessment may improve this rating, but success not clearly expected based on available information.
- **UNTESTED.** Efficacy, reliability and repeatability not established due to lack of or insufficiency of reported testing. Commercials and testimonials do not count as documentation of positive outcomes.

This matrix is used as a quick reference for what works or not:

APPLICABILITY \ RELIABILITY	HIGH	MEDIUM	LOW	UNTESTED
HIGH	Works	Probably works	May work (beware)	Not recommended
MEDIUM	Probably works	May work (beware)	Unlikely to work	Not recommended
LOW	Not recommended	Not recommended	Not recommended	Not recommended

The meaning of these classifications is:

- **WORKS.** When applied in appropriate situations, this approach or technique provides predictable and measurable outcomes with a high level of reliability.
- **PROBABLY WORKS.** When applied in appropriate situations, this approach or technique provides predictable and measurable outcomes with a reasonable level of reliability.
- **MAY WORK (BEWARE).** When applied in appropriate situations, this approach or technique may provide measurable outcomes, albeit with a low level of reliability. Caution is advised.
- **UNLIKELY TO WORK.** This approach or technique is not likely to provide reliable positive outcomes.
- **NOT RECOMMENDED.** This approach or technique cannot be recommended because it has unknown efficacy or it has not been tested (or both). Techniques with otherwise "high" applicability may be considered for experiment evaluation.

In addition, we have included ratings for the duration of benefits, maintenance requirements and costs, as follows:

DURATION OF BENEFITS

How long can the benefits be expected to last?

- **LONG.** Multiple seasons or years.
- **MEDIUM.** About a season or year.
- **SHORT.** Less than a month.

APPLICATION REQUIREMENTS

We are unaware of any technique that lasts forever. Some require ongoing operation, some need reapplication infrequently and some in between.

- **CONTINUOUS.** Continuous operation or application required for benefits.
- **FREQUENT.** Application required once or more per season.

- **SEASONAL.** Application required no more than once each season.
- **OCCASIONAL.** Periodic reapplications are required to maintain benefits, but less than once per year.
- **RARE.** Applications should last for many years.

COSTS

There typically is a wide range of costs for managing algae problems. Here we have tried to normalize these costs by lake acreage and amortizing them on an average annual basis. Thus, costs are presented in wide ranges as $ / acre / year.

- **LOW.** Less than $500/acre/year.
- **MEDIUM.** $500 to $1,000/acre/year.
- **HIGH.** Greater than $1,000/acre/year.

BEST PRACTICES FOR ALGAE PROBLEMS

Any algae control technique, technology or tool may be more or less appropriate for a particular situation. The guidance we present is intended as an aide for critical evaluation. This information should help guide you in the right direction.

For each technique, technology or tool, we provide an assessment of its overall rating as well as its applicability, reliability, duration of effectiveness, application requirements and cost. Our assessments are listed alphabetically.

ALGAECIDES

Algaecides are chemicals that kill algae and have been in use for many years. The most commonly used algaecides are copper compounds or hydrogen peroxide. Algaecides' efficacy tends to be short-lived and there are cases where algae, especially blue-green algae have become resistant to algaecide treatments. The duration of effectiveness is short (weeks) and frequent repeated applications are typically indicated. The cost varies widely, depending on the number of treatments per season (typical range is 1 – 5, which accounts for the "low" to "medium" cost range). Normally, large portions of the lake are treated, unless there are isolated bays with problems.

LAKE ADVOCATES' RECOMMENDATION: Algaecides have predictable outcomes and "work" in applicable situations. Algaecides should be viewed as a short-term remedy to algae problems. We do not recommend algaecides as a long-term solution and their negative ecological effects must also be considered.

ALGAECIDES	
WORKS	
APPLICABILITY	**HIGH.** Applicable to stated problem
RELIABILITY	**HIGH.** Positive outcomes well established
DURATION	**SHORT.** Less than a month
APPLICATION	**FREQUENT–SEASONAL.** Once, to once or more per season
COST	**MEDIUM-LOW.** Up to $500/acre/year

ARTIFICIAL CIRCULATION

Artificial circulation uses machines, usually bubblers or circulators, to extend the depth or duration of water circulation. Artificial circulation, implemented critically, is not for the timid – it requires serious engineering, appropriate equipment, adequate power and comes with significant costs. To be effective, artificial circulation has to address light availability and the rate of circulation to inhibit photosynthesis. Specific design steps to meet these demands are needed. When applied uncritically or lacking adequate diagnostics, engineering, or power, artificial circulation is neither reliable nor applicable and can sometimes do harm. Uncritically applied artificial circulation can be tempting, especially when it appears to be cheap. There are many cases where artificial circulation is misapplied or applied to inappropriate problems. In many of these cases, this technique is recommended based mainly on testimonials.

People like seeing bubbles in their lakes.

There have even been claims that artificial circulation will control rooted or invasive plants - we are not aware of any cases where artificial circulation has been demonstrated to be effective at controlling plants. Below, reliability and applicability are "high" in cases where artificial circulation is applied critically and appropriately, but "low" in cases where artificial circulation is applied uncritically.

LAKE ADVOCATES' RECOMMENDATION: Artificial circulation is a reliable algae management technique when indicated as appropriate and implemented correctly. When applied uncritically, artificial circulation is unreliable and therefore, not recommended. We caution against considering "to good to be true" claims and assurances with ready-made bubblers, fountains and similar devices. Similarly, we caution when considering aeration as enhanced by microbes or enzymes (see the following).

ARTIFICIAL CIRCULATION (CRITICAL)

WORKS

APPLICABILITY	**HIGH.** Applicable to stated problem
RELIABILITY	**HIGH.** Positive outcomes well established
DURATION	**SHORT.** Multiple seasons or years
APPLICATION	**CONTINUOUS.**
COST	**HIGH-MEDIUM.** $500/acre/year or more

ARTIFICIAL CIRCULATION (UNCRITICAL)

NOT RECOMMENDED

APPLICABILITY	**LOW.** Not applicable to stated problem
RELIABILITY	**LOW.** Efficacy, reliability, and repeatability not clearly demonstrated
DURATION	**N/A.**
APPLICATION	**N/A.**
COST	**N/A.**

BIOMANIPULATION

Biomanipulation refers to manipulations aimed at multiple links in the food chain. Manipulations can involve physical alterations or the stocking of predators or herbivores. The efficacy, reliability and repeatability of biomanipulation increases with decreased phosphorus levels in lakes; that is, it is most effective when it may be least required. There are documented cases where biomanipulation works well, however these involve ongoing maintenance and inputs of energy. Cases involving long-term evaluations are substantially lacking. We therefore evaluate two categories: Biomanipulation as ongoing, long-term programs involving intensive monitoring, evaluation and adjustment and biomanipulation attempted as a one-time or short-term manipulation lacking critical assessment.

LAKE ADVOCATES' RECOMMENDATION: Biomanipulation requires active, ongoing monitoring, modification inputs of energy. Shortcuts in these areas are not recommended. We also recommend that phosphorus reductions be included in a long-term management strategy, as this will better sustain biomanipulation, ideally making it unnecessary in the long run, by relegating it to a supporting method for enhancing lake uses, like fishing.

BIOMANIPULATION (LONG-TERM)

MAY WORK (BEWARE)

APPLICABILITY	**MEDIUM.**
RELIABILITY	**HIGH.** Positive outcomes well established
DURATION	**MEDIUM.** About a season or a year
APPLICATION	**OCCASIONAL.** Periodic reapplications required
COST	**LOW-HIGH.**

BIOMANIPULATION (SHORT-TERM)

NOT RECOMMENDED

APPLICABILITY	**LOW.** Not applicable to stated problem
RELIABILITY	**LOW.** Efficacy, reliability and repeatability not clearly demonstrated
DURATION	**MEDIUM.** About a season or a year
APPLICATION	**SEASONAL.** Application no more than once per season
COST	**MEDIUM-LOW.** Up to $500/acre/year

DRAWDOWN

Lake drawdowns require the ability to drain large volumes of water for extended periods of time. This is impractical in many cases or may have unacceptable non-target impacts, so acceptability may be low. When accomplished, drawdown exposes shallow sediments thereby desiccating or freezing plants (which will not occur in warmer climes). Rooted plant control is typically the primary target of drawdown – phosphorus control then may be an incidental benefit through sediment modification. Many rooted plants will be controlled for multiple seasons however, plant species that germinate from seeds annually may increase following a drawdown. In cases where phosphorus is mobilized from shallow sediments, drawdown may also mitigate internal phosphorus inputs, but decomposition of sediment may release additional nutrients. Costs vary.

This description of drawdown is simplistic. Numerous factors and considerations go into the diagnostic and design – well beyond the scope provided here.

LAKE ADVOCATES' RECOMMENDATION: Drawdown should be considered when appropriate diagnostic studies demonstrate a phosphorus reduction benefit.

DRAWDOWN

MAY WORK (BEWARE)

APPLICABILITY	**MEDIUM**
RELIABILITY	**MEDIUM.** Positive outcomes likely
DURATION	**MEDIUM.** About a season or a year
APPLICATION	**OCCASIONAL.** Periodic reapplications required
COST	**LOW-HIGH.**

DREDGING

Dredging and removal of lake sediments has multiple benefits, including the removal of nutrient-enriched sediments and deepening the lake. As a result, internal phosphorus recycling may be diminished. Disposal of dredged spoils is often an onerous issue, especially if they contain hazardous materials. While dredging may be considered perhaps the only true lake restoration technique, it is, unfortunately, very expensive and therefore not always feasible. In most cases, a Phosphorus Inactivation treatment (such as alum) following the dredging is recommended or required (see below).

LAKE ADVOCATES' RECOMMENDATION: Dredging and removal of nutrient-laden sediments is a truly ecological remedy, when feasible and affordable.

DREDGING	
PROBABLY WORKS	
APPLICABILITY	**HIGH.** Applicable to stated problem
RELIABILITY	**MEDIUM.** Positive outcomes likely
DURATION	**LONG.** Multiple seasons or years
APPLICATION	**RARE.** Applications last for many years
COST	**HIGH.** More than $500/acre/year

MICROBES AND ENZYMES

Microbes, bacterial concoctions or enzyme treatments, sometimes augmented with artificial circulation, promise to facilitate algae control or nutrient manipulations. We are aware of no objective documentation of positive outcomes. Most claims of efficacy lack objective, third-party evaluations - claims of efficacy usually originate from the vendors.

There is sound theoretical underpinning for microbe use based on wastewater management, but practical application in lakes lacks scientific support. Far more third-party monitoring and evaluation is needed than is offered by available products.

LAKE ADVOCATES' RECOMMENDATION: Microbes and enzymes lack objective evaluations, therefore their use for algae control is not recommended.

MICROBES & ENZYMES	
NOT RECOMMENDED	
APPLICABILITY	**LOW.** Not applicable to stated problem
RELIABILITY	**UNTESTED.**
DURATION	**N/A.**
APPLICATION	**N/A.**
COST	**N/A.**

OXYGENATION

Oxygenation is a kind of aeration that adds concentrated oxygen to the lake water, which provides increased habitat and limits (but does not always eliminate) internal phosphorus recycling. Oxygen is most often added below the thermocline, called hypolimnetic aeration. Oxygenation requires specialized equipment, careful planning, proper design and precise implementation. Capital expense is high and operational expenses are variable. When low dissolved oxygen facilitates internal P recycling, algae problems can be mitigated by adding oxygen to deeper water.

LAKE ADVOCATES' RECOMMENDATION: When indicated as appropriate, oxygenation can be effective.

OXYGENATION

WORKS

APPLICABILITY	**HIGH.** Applicable to stated problem
RELIABILITY	**HIGH.** Positive outcomes well established
DURATION	**SHORT.** Multiple seasons or years
APPLICATION	**CONTINUOUS.**
COST	**HIGH-MEDIUM.** $500/acre/year or more

PUBLIC EDUCATION

Public education[5] is perhaps the most recommended, most used management approach and it is also the least objectively studied with respect to physical outcomes, such as mitigating algae problems. The reliability of public education is "low" in the context of evaluating measurable, tangible outcomes of

---

[5] Here we discuss public education in the context of discussing "what works" specifically regarding measurable outcomes for addressing algae problems. Public education is relevant and important in the broader context of the overall lake management program.

direct results or lake condition. Similarly, the expectations of positive outcomes are substantially unknown due to the lack of objective evaluation. For the most part, public education is applied uncritically – it is viewed as the right thing to do, often considered sufficient on its own to mitigate problems, but least documented in terms of actual results. More studies by social scientists are showing it is possible under some circumstances to change attitudes and sometimes behaviors, however we are aware of no documented beneficial outcomes in terms of lake condition or positive departures from a known baseline condition based sole upon public education.

Public education undoubtedly has merits, but is insufficient to recommend with the expectation for measurable outcomes in lakes.

Public education may be used to reduce human health risks by informing the public about exposure to harmful algal blooms.

LAKE ADVOCATES' **RECOMMENDATION:** Public education has obvious merits for raising awareness and therefore should be included in lake management programs. We caution against relying on public education programs alone to mitigate algae problems, **Lake Advocates** recommends that public education and public engagement be part of a lake management program, but not the main method of P or algae control.

PUBLIC EDUCATION

NOT RECOMMENDED

APPLICABILITY	**LOW.** Not applicable to stated problem
RELIABILITY	**LOW.** Efficacy, reliability, and repeatability not clearly demonstrated
DURATION	**N/A.**
APPLICATION	**N/A.**
COST	**N/A.**

* Not recommended as a reliable method for mitigating algae problems, but public education should be included as an element supporting a sustainable lake management program and protecting human health.

WATERSHED MANAGEMENT - PREVENTION

In cases where algae problems have not yet occurred, prevention is a rational and responsible strategy. Land development threatens to upset this balance and often leads to phosphorus impairments and concomitant algae problems. Successful cases demonstrating effective prevention programs involve small watersheds (less than 10-times the lake's surface area), a plan with identified priority areas, an organized and committed community, a reliable funding source, the ability to protect critical lands through outright purchase, conservation easement or other similar practice and ongoing field monitoring.

LAKE ADVOCATES' RECOMMENDATION: Investments in preventing algae problems through watershed protection for the long-run, while involving serious commitment and expense, are rational and responsible.

WATERSHED - PREVENTION	
PROBABLY WORKS	
APPLICABILITY	**HIGH.** Applicable to stated problem
RELIABILITY	**MEDIUM.** Positive outcomes likely
DURATION	**LONG.** Multiple seasons or years
APPLICATION	**OCCASIONAL.** Periodic reapplications required
COST	**HIGH.** More than $500/acre/year

WATERSHED MANAGEMENT – BEST MANAGEMENT PRACTICES

Watershed management that relies on best management practices (BMPs, for example, rain gardens, detention ponding, street sweeping) is the predominant watershed management paradigm. This approach is based on the presumption that the ultimate source of phosphorus is delivered from a lake's watershed. While this is strictly accurate, it often does not follow that reducing these sources will lead to improvements in lake condition. In addition, the phosphorus reduction of many BMPs is insufficient relative to P loading to the lake.

Mitigating phosphorus impairments requires reductions in phosphorus inputs of 80% (sometimes more). While some individual BMPs may reduce phosphorus by up to 50%, only up to about 25% is likely on a watershed-wide basis – thus the insufficiency.

Unfortunately, phosphorus runoff in developed watersheds is hard-wired and BMPs are not up to the job of reversing this. There is no way to make runoff from developed lands mimic runoff from undeveloped lands.

In addition to the insufficiency of watershed management using BMPs, many impaired lakes also have issues with internal phosphorus loading, which diminishes the lake's resilience and retards the lake's recovery.

The percentage of documented watershed projects that have mitigated known phosphorus impairments using BMPs alone, even those involving decades and substantial public costs, is extremely low.

LAKE ADVOCATES' RECOMMENDATION: Most lake management programs rely on or even require the implementation of BMPs to mitigate excess phosphorus inputs to lakes. Unless this strategy can be demonstrated through site-specific monitoring, modeling and verification to be sufficient, we think it is unlikely to work. We also caution that once a lake is experiencing algal blooms, watershed activities alone will not fix the in-lake problem in a reasonable amount of time, that is, decades or more, if ever.

Although watershed BMPs alone will not likely restore a lake that is impaired and has significant internal P loading, *Lake Advocates* recommends that watershed management be part of a long-term implementation program.

WATERSHED - BMP	
UNLIKELY TO WORK	
APPLICABILITY	**MEDIUM.**
RELIABILITY	**LOW.** Efficacy, reliability, and repeatability not clearly demonstrated
DURATION	**N/A.**
APPLICATION	**N/A.**
COST	**N/A.** (although costs are likely high)

WATERSHED MANAGEMENT – END-OF-PIPE

A promising area of emerging technology involves engineering or chemical approaches that capture and treat phosphorus in runoff before it enters a lake. There are cases where phosphorus reductions from entire drainages are reduced by 90%, sufficient for beneficial impacts. These approaches can be costly, but are usually far less costly than a BMP approach. And the benefits are realized almost immediately.

LAKE ADVOCATES' **RECOMMENDATION:** In cases where meaningful, measurable and timely improvements are desired (or required), end-of-pipe approaches are often the most effective watershed management approach.

WATERSHED - END-OF-PIPE

PROBABLY WORKS

APPLICABILITY	**HIGH.** Applicable to stated problem
RELIABILITY	**MEDIUM.** Positive outcomes likely
DURATION	**LONG.** Multiple seasons or years
APPLICATION	**CONTINUOUS.**
COST	**HIGH-MEDIUM.** $500/acre/year or more

PHOSPHORUS PRECIPITANTS – ALUM AND ALUMINUM COMPOUNDS

Aluminum compounds, especially aluminum sulfate (alum), have a long documented record of effective treatments to mitigate excess phosphorus through one of these approaches: phosphorus water column stripping, phosphorus sediment inactivation, phosphorus interception and phosphorus maintenance. Early concerns with toxicity have been solved with appropriate implementation. Dose calculations must be done properly, but success is undeniable. Duration and maintenance vary depending on application strategy. Treatment costs may appear high, but are commonly much cheaper per pound of phosphorus mitigated than other methods.

It is a common misconception that aluminum is a one-time treatment for sealing the bottom sediments to retard phosphorus recycling. Aluminum treatment strategies are more diverse and can be used to target various aspects of a lake's phosphorus cycle.

WATER COLUMN STRIPPING involves a low dose surface application intended to strip phosphorus from the water. The objective is to accelerate lake recovery following other phosphorus mitigation efforts. The duration of effectiveness is intended to be short.

PHOSPHORUS INACTIVATION involves larger doses applied to the surface, but intended to retard internal phosphorus cycling. When the appropriate dose is applied, the longevity of effectiveness is up to about 10 years for shallow lakes and 13 to 30 years for deep lakes. Re-applications are expected to maintain efficacy.

PHOSPHORUS INTERCEPTION involves intercepting runoff, treating with alum, then returning the treated water to the runoff stream. This strategy is a common end-of-pipe watershed management approach (see above).

ALUMINUM FOR PHOSPHORUS MAINTENANCE is used to periodically treat lake water or sediments to maintain lake phosphorus concentration targets.

***LAKE ADVOCATES'* RECOMMENDATION:** Aluminum to mitigate excess phosphorus from various sources is a reliable tool, especially when the appropriate strategy is used, the dose is calculated correctly and it is applied properly.

P PRECIPITANTS - ALUM	
WORKS	
APPLICABILITY	**HIGH.** Applicable to stated problem
RELIABILITY	**HIGH.** Positive outcomes well established
DURATION	**VARIABLE.**
APPLICATION	**VARIABLE.**
COST	**MEDIUM-LOW.** Up to $500/acre/year

PHOSPHORUS PRECIPITANTS – CALCIUM & IRON

Phosphorus precipitation using other metal salts, particularly calcium (Ca) or iron (Fe), has a long documented record of effective treatments to mitigate excess phosphorus in appropriate situations. The biggest drawbacks are effects of pH and oxygen, which limit applicability. Special handling and application methods are typically required. Where applicable, positive outcomes are reliable.

LAKE ADVOCATES' RECOMMENDATION: When applicable, calcium or iron compounds can be effective phosphorus precipitants.

P PRECIPITANTS - CA, FE	
PROBABLY WORKS	
APPLICABILITY	**HIGH.** Applicable to stated problem
RELIABILITY	**HIGH.** Positive outcomes well established
DURATION	**VARIABLE.**
APPLICATION	**VARIABLE.**
COST	**MEDIUM-LOW.** Up to $500/acre/year

PHOSPHORUS PRECIPITANTS – OTHER

There are newer products available with claims they are effective phosphorus precipitants. Examples include Baraclear™ (aluminum sulfate, sodium bentonite and calcium carbonate), Phoslock™ (lanthanum) and SeClear™ (algaecide with unspecified metal salt). We are aware of no peer-reviewed (except lanthanum, see next paragraph), third party evaluations or field trials, and the track record is too limited to draw conclusions.

More recently, there has been a large number of peer-reviewed articles regarding the use of lanthanum. Many of these cases show improved water quality, at least in the short-term. There is an overall lack of data on the longevity of lanthanum, so for now, we assume it is best used as a phosphorus stripping strategy. Similarly, there is very little cost data available, but some studies indicated a substantially greater cost (compared to alum); for example, at least 5- to 100-times more (for example, Black Lake, WA).

With proper evaluation, some of these products could be reliable where appropriate, although costs may be high.

LAKE ADVOCATES' **RECOMMENDATION:** We do not recommend the use of other precipitants at this time.

P PRECIPITANTS - OTHER

NOT RECOMMENDED

APPLICABILITY	**MEDIUM.**
RELIABILITY	**UNTESTED.**
DURATION	**N/A.**
APPLICATION	**N/A.**
COST	**N/A.**

DISCUSSION

The algae control best practices screening tool, presented here, should not be used to make final decisions. Methods that come through as "works," mean they should be further considered and evaluated. Methods listed as "probably," "may," etc. may still be considered, but greater care should be exercised as indicated in the definitions presented here.

Further, it is essential to follow good planning steps. These include problem definition, diagnostics, modeling and feasibility assessment, setting measurable objectives, monitoring and evaluation.

Of course, a management tool or technique that "works" does not necessarily mean it should be used. A hammer "works," but should not be used for a screw.

Economic and institutional factors may also limit application of what works, but use of alternatives for cost or permitting reasons does not change their status in this evaluation.

Some management techniques that do not neatly fit in larger categories are not considered (for example, barley straw or floating islands). Generally, techniques not included above have insufficient evaluation to warrant recommending their use.

There are a number of reliable, tried-and-true methods available for managing algae problems. Many of these have become out of vogue for various reasons. Many contemporary lake management approaches are being tried because they are viewed as correct (for example, public education, watershed management), they are sexy or natural (for example, microbes, enzymes), they are non-chemical (for example, bubblers or solar circulators) or they appear innovative. But these are largely untested (and therefore unreliable) or ineffective.

Our profession has gotten away from government or third party demonstrations, which has put us in a poor position to evaluate new technologies. In this vacuum, new products have emerged to fill perceived needs, with vendors and manufacturers providing self-evaluations. We have seen too many projects that have wasted time and money on techniques that yield minimal or no results.

Lack of federal, state and often local funding has forced a desperate public to consider or use lake management techniques that are affordable. Unfortunately, these techniques are also often ineffective. Our current institutional approach emphasizes watershed management, has spent or caused to have spent hundreds of millions of dollars, yet too many lakes remain impaired. And who doesn't like public education? Nevertheless there is a lack of field studies demonstrating positive results in our lakes.

We hope and intend for this assessment to be useful for those wanting to improve their lakes to help narrow and focus their efforts.

Please contact *Lake Advocates*
for more information or assistance.

www.LakeAdvocates.org

GLOSSARY

ABUNDANCE. An ecological concept referring to the relative representation of a species in a particular ecosystem. It is usually measured as the number of individuals per volume or area within a sample.

ALGAE. Non-vascular aquatic plants that occur as single cells, colonies, or filaments. They contain chlorophyll, but lack special water-carrying tissues. Through the process of photosynthesis, algae produce food and oxygen in water environments.

ALGAL BLOOM. Rapid growth of algae populations with increase in biomass in response to eutrophic conditions (over enrichment of nutrients).

ALUM[2]. Aluminum sulfate, an aluminum salt, that is used to lower lake P content by removing phosphorus in the water (through chemical precipitation) and by retarding release of mobile P from lake sediments (P inactivation). Alum (aluminum sulfate) is added to the water column to form aluminum phosphate and a colloidal aluminum hydroxide floc to which certain P fractions are bound. The aluminum hydroxide floc settles to the sediment and continues to sorb and retain P.

ANOXIC. This condition exists when there is no oxygen in the water or interstitial sediment water. Commonly characterized by very low concentrations of dissolved oxygen (< 2 mg/L DO).

BIOMASS. The weight of biological matter. Standing crop is the amount of biomass (e.g., fish or algae) in a body of water at a given time. Often measured in terms of grams per square meter of surface area or volume of water.

CHLOROPHYLL. A green pigment in algae and other green plants that is essential for the conversion of sunlight, carbon dioxide, and water to sugar. Chlorophyll present in all types of algae, sometimes in direct proportion to the biomass of algae.

EUTROPHIC. A eutrophic lake is rich in nutrients and organic materials. Excessive loading of plant nutrients, organic matter, and silt cause increased primary producer (e.g. algae) biomass, reduced water clarity, good growing conditions for nuisance species, and usually decreased lake volumes over time. Synonymous with phosphorus-impairment.

Oligotrophic is the opposite of eutrophic – nutrient-poor. And mesotrophic is in between.

EUTROPHICATION. The process of physical, chemical, and biological changes associated with nutrient, organic matter and silt enrichment and sedimentation of a lake that cause a water body to age and become eutrophic. Symptoms can include dissolved oxygen depletions and fish kills due to over production of plants and algae.

EXTERNAL LOADING. The total amount of material (sediment, nutrients, oxygen-demanding material) brought into the lake by inflowing streams, runoff, direct discharge through pipes, groundwater, the air, and other sources over a specific period of time.

INTERNAL NUTRIENT LOADING. The total amount of nutrients released into the water column over a specific period of time as a result of nutrient recycling from sediments, wind resuspension, mineralization of nutrients, macrophyte senescence, and decomposition of organic material.

LOADING. See external loading and internal nutrient loading.

NON-POINT SOURCE. Pollution that cannot be traced to specific origin or starting point, but seems to flow from many different sources. Non-point source pollutants are generally carried off the land by storm-water runoff.

NUTRIENT. An element or chemical essential to life, such as carbon, oxygen, nitrogen, and phosphorus. Nutrients promote growth and repair the natural destruction of organic life.

PHOSPHORUS-IMPAIRMENT. There is substantial practical and programmatic overlap in the definition (in the US) of "eutrophic" and "P-impaired" lakes. Eutrophication is a more general ecological term and impairment has specific references to the administration of the Clean Water Act. Here eutrophic is used synonymously with and inclusive of P- impairment.

PHYTOPLANKTON. Microscopic algae and microbes that float freely in open water of lakes. In some lakes, they provide the primary base of the food chain for all animals. They also produce oxygen by a process called photosynthesis. This usually includes cyanobacteria (formally called cyanophyta or bluegreen algae) that are real bacteria and not green plants, but utilize the ecological niche and are photosynthetic using chlorophyll a.

POINT SOURCE. Pollution discharged into water bodies from specific, identifiable pipes or points, such as an industrial facility or municipal sewage treatment plant.

RUNOFF. That portion of precipitation that flows over the land carrying with it such substances as soil, oil, trash, and other particulate and dissolved materials until it ultimately reaches streams, rivers, lakes, or other water bodies.

SECCHI DEPTH. A measure of transparency of water (the ability of light to penetrate water) obtained by lowering a black and white disk (Secchi disk, 8-inches in diameter) into water until it is no longer visible. Measure in units of meters or feet.

SEDIMENT. Bottom material in a lake that has been deposited after the formation of a lake basin. It originates from remains of aquatic organisms, chemical precipitation of dissolved minerals, and erosion of surrounding lands.

STRATIFICATION. Process in which several horizontal water layers of different density may form in some lakes. During stratification, the bottom mass (hypolimnion) is cool, high in nutrients, low in light, low in productivity, and low in dissolved oxygen. The top mass (epilimnion) is warm, higher in dissolved oxygen, light, and production, but lower (normally) in nutrients. The sharp boundary between the two masses is called a thermocline that is usually found within the transition layer of water called the metalimnion.

THERMOCLINE. A transitional layer between the epilimnion and the hypolimnion;the rate of temperature change with depth is greatest in this layer.

WATER COLUMN. Water in the lake between the interface with the atmosphere at the surface and the interface with the sediment layer at the bottom.

WATERSHED. A drainage area or basin in which all land and water areas drain or flow toward a central collector such as a stream, river, or lake at a lower elevation.

ABOUT THE AUTHORS

DICK OSGOOD

Dick Osgood is educated as a scientist (MS Aquatic Ecology & Geology; BS Biology), experienced as an Environmental Planner and is a Certified Lake Manager (one of only 75 in the world), and has worked for and with public, private and nonprofit organizations. Dick specializes in developing lake management plans, invasive species management, diagnostic studies, modeling and alum dosing. Dick has authored numerous scientific journal papers, made hundreds of presentations at professional meetings, is the author of regular columns and has frequently served as an expert witness. Dick is also trained as a mediator and facilitator. Dick has served on the Board of Directors of the North American Lake Management Society (Past-President), Minnesota Waters and Minnesota Lakes Association (Officer & Public Policy Committee chair), Conservation Minnesota, Excelsior Rotary Board and the South Lake-Excelsior Chamber of Commerce Board. Dick was an invited delegate to the Symposium on the Ecological Basis for Lake and Reservoir Management at the University of Leicester, England, is a co-instructor of the Alum Workshop at the Annual North American Lake Management Society, has authored a chapter in the book 'Managing Lakes and Reservoirs', and Dick's consulting business, Osgood Consulting, was named the North American Lake Management Society's 'Outstanding Corporation' in 2005.

HARRY GIBBONS

Dr. Gibbons has 40 years of experience in applied limnology, lake, reservoir, river, stream, and wetland restoration. Harry has specifically planned/designed management and restoration programs for over 280 lakes/reservoir, over 100 ponds, and 40 river systems. Harry earned his Ph.D. in limnology and MS in

Environmental Engineering from Washington State University and his BS in biology from Gonzaga University. His expertise includes lake and watershed management, lake restoration, integrated aquatic plant management, aquatic invasive species (AIS) management (developing non-point source control solutions), stream assessment, fish passage, aquatic habitat assessment, wetland restoration and storm-water management. Harry is a recognized leader in the development and implementation of in-lake activities for techniques like phosphorus inactivation (alum), dredging, hypolimnetic aeration, aeration and complete circulation, AIS management, and integrated aquatic plant management. His in-lake and wetland BMP designs have earned him environmental excellence awards, for example Phantom/Larsen Lakes and Lake Stevens restoration have received award honors from the Consulting Engineering Council of Washington for excellence in environmental projects in 1993 and 1995. Both projects included hypolimnetic aeration, stream and wetland restoration, storm-water BMPs and habitat enhancement. In 2012, Harry was given the Secchi Disk Award by the North American Lake Management Society for his outstanding technical contributions and service to help promote lake management. Harry has served on the North American Lake Management Society Board of Directors three times, 1992-1994, 2004-2006, and 2008-2010 as President Elect, President and Past President. He was a founding member and served on the board of the Washington State Lake Protection Association and was President in 1990. Harry has been a research graduate faculty at Washington State University (1981-1984), and Portland State University, (1998-1999), and for over 30 years the lead limnologist for Tetra Tech, Inc. In addition to his lake work, he has conducted comprehensive river and reservoir limnological studies in several major river systems including the Columbia, Snake, Spokane, Clearwater, Chehalis, Green, Wynoochee, Susitna and Pend Oreille Rivers, including 15 hydroelectric reservoirs. Harry has authored numerous technical reports and limnological journal and magazine articles, he was also co-presenter and author for the *Phosphorus Inactivation & Interception*

Workshop & Manual, 2002-15, co-author of *A Citizen's Manual for Developing an Integrated Aquatic Vegetation Management Plan* for the Washington State Department of Ecology, NALMS' *Aquatic Plant Management in Lakes and Reservoirs,* 1996 for EPA, and *Guide for Developing Integrated Aquatic Vegetation Management Plans in Oregon,* Portland State University, 1999.

Figure 6. Dick and Harry (Dick is the one without hair color, but Harry is the old man).